CURTISS-WRIGHT CW-21

INTERCEPTOR

CW-21 INTERCEPTOR

This sleek looking CW-21 with civil registry NX19441 was one of three destined for China. It first flew on 20 March 1940.

(collection E. Hoogschagen)

Previous page:
A publicity photo dated 1 December 1940. The CW-21 was pitched as the mile a minute fighter, the ideal fighter for intercepting high flying enemy bombers. Seen here is a CW-21B prior to delivery to the Netherlands Indies.

(collection M. Willis)

INTRODUCTION

During the early stages of the Sino-Japanese war, which had erupted in July 1937, it quickly became apparent that the Chinese air force urgently required modern fighter aircraft which combined a high rate of climb with superior manoeuvrability.

The Central Aircraft Manufacturing Company (CAMCO), founded by American William Douglas Pawley, had assembled a series of Curtiss Hawk II and III fighter planes for the Chinese Nationalist government. Pawley was also president of Intercontinental Corporation, this company was sales representative for Curtiss-Wright in China during the second half of the 1930s. The strong ties between Pawley and Curtiss-Wright were the stepping stone for development of a new light weight fighter which would meet the demands of modern air war.

The CW-21B featured an inward retracting landing gear and other refinements. It first flew mid–September 1940.

(collection J. Grisnich)

DEVELOPMENT

Work on this new fighter type started in the St. Louis branch of the Curtiss-Wright Company. The most important design philosophy was to employ a high power engine combined with an airframe as light and compact as possible. The design team, led by Willis Wells (project engineer) and George A. Page jr. (chief engineer), drafted a design which would evolve into the CW-21 fighter.

Right:
The Curtiss-Wright CW-19L was the company's first all metal aircraft design and paved the way to other designs, including the CW-20 transport and CW-21 fighter.

(collection N. Braas)

Curtiss-Wright chief designer George A. Page jr., in front of a CW-25, the AT-9 Jeep advanced trainer.

(Missouri historical society collection)

THE CURTISS-WRIGHT COMPANY

The Curtiss-Wright company was formed on 26 June 1929 when the Curtiss Aeroplane and Motor Company merged with the Wright Aeronautical Corporation. The company was organised in an aircraft division (Curtiss) and a division constructing engines and propellers (Wright). The headquarters was located in New York City. During the years before the merger Curtiss had purchased the Keystone, Loening, Robertson, Travel Air and Moth companies, but these were all dissolved during the Great Depression of the 1930s. Production of some types were continued under the Curtiss-Wright company name at the St. Louis facilities – the former Robertson plant. Other businesses such as the Curtiss-Wright Flying service had to discontinue activities.

Above:
An aerial view of the Curtiss-Wright facilities at Lambert field, St. Louis. (Missouri historical society collection)

Bottom:
The Curtiss-Wright complex after completion of expansion work, December 1941. (Missouri historical society collection)

The completely redesigned CW-19R-12 advanced trainer, armed with two forward firing light machineguns.
(Missouri historical society collection)

The new fighter shared some design elements first used in the CW-19L, which was originally designed by the Curtiss-Robertson company as CR-2 Coupe, before this company was absorbed in the Curtiss-Wright company. The CW-19L was introduced in 1935 and was intended as private aircraft featuring side by side seats and streamlined fixed landing gear, covered with trouser type fairings. Although the CW-19L had unfavourable stall characteristics and a tendency to ground loop, it did have an excellent rate of climb – it reached 23,000 feet in 15 minutes. Many of its vices were solved after it received a completely redesigned wing. The design showed potential and it was redeveloped as tandem seater, intended for military use. This variant was known as CW-19R. Twenty were sold to China and small quantities were bought by Bolivia, Cuba, Dominican Republic and Ecuador.

The unarmed CW-19R advanced trainer in Ecuadorian markings. (collection E. Hoogschagen)

The completed CW-21 prototype, registered NX19431, made its first flight at Wright Field on 27 September 1938, piloted by Ned Warren. At this time, the US Army Air Corps showed no interest in the plane.

TECHNICAL DESCRIPTION

The CW-21 featured a very slim fuselage design with highly tapering tail section aft of the cockpit. The overall layout of wings and tail surfaces were similar to the CW-19R, although the wing surfaces for CW-21 were strengthened.

The monocoque fuselage was entirely made of aluminium. Armament – a pair of .30 machineguns with 1000 rounds each – was fitted behind the firewall and fired through the propeller arc. The pilot sat in a fully enclosed cockpit with rearward sliding canopy. A bulkhead constructed out of aluminium plate directly behind the pilot served as turnover column. The front was made out of ¼ In (6.3 mm) steel sheet and provided a limited armour protection for the pilot. The central wing section was bolted to the fuselage and held two fuel tanks of 34 gallons each. The main landing gear legs retracted rearwards into fairings

The all-green CW-21 prototype NX19431 in flight. (collection M. Willis)

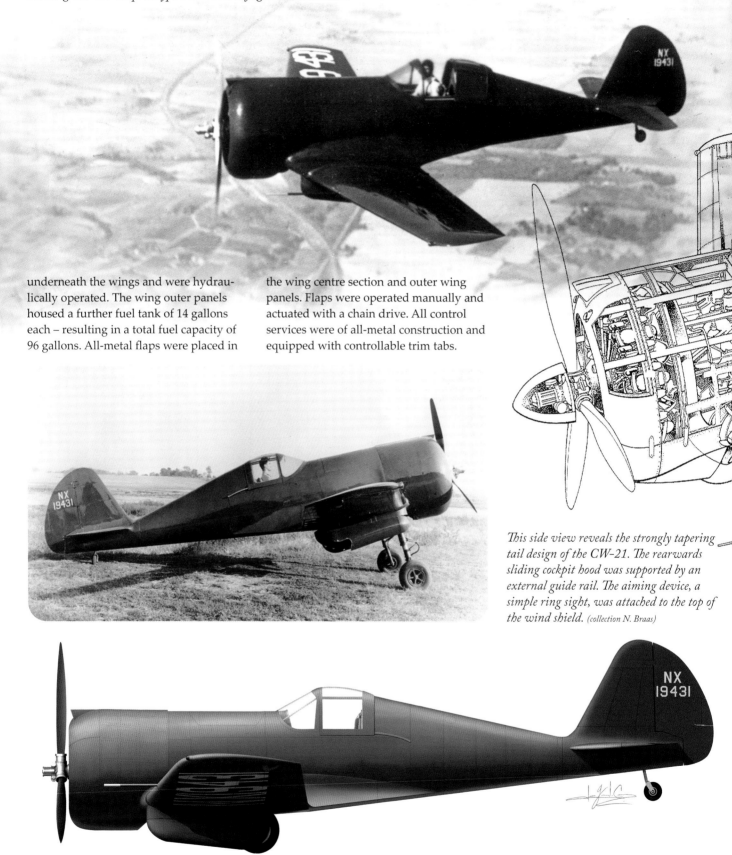

underneath the wings and were hydrau-lically operated. The wing outer panels housed a further fuel tank of 14 gallons each – resulting in a total fuel capacity of 96 gallons. All-metal flaps were placed in the wing centre section and outer wing panels. Flaps were operated manually and actuated with a chain drive. All control services were of all-metal construction and equipped with controllable trim tabs.

This side view reveals the strongly tapering tail design of the CW-21. The rearwards sliding cockpit hood was supported by an external guide rail. The aiming device, a simple ring sight, was attached to the top of the wind shield. (collection N. Braas)

The CW-21 prototype with registry NX19431. It was painted in an overall coat of green, most likely Dark Green shade 30. Profile art by Luca Canossa.

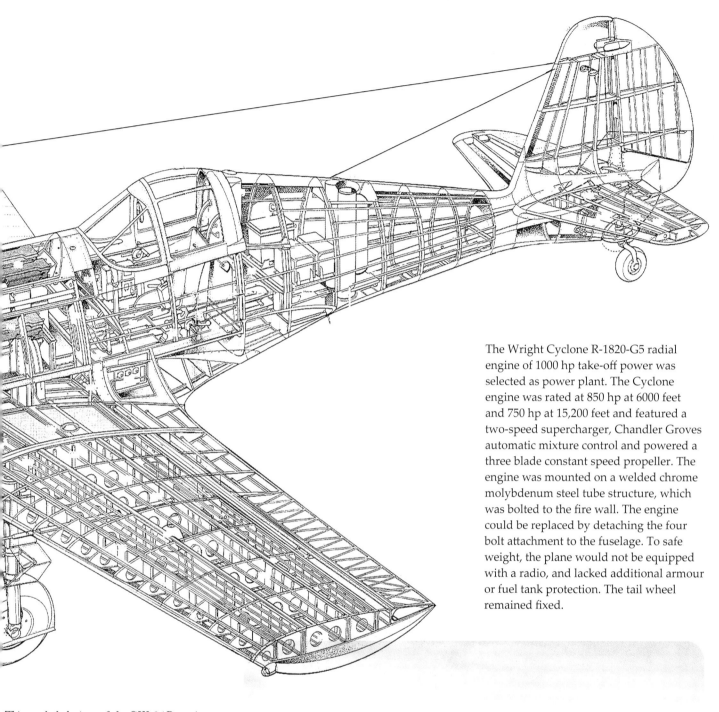

The Wright Cyclone R-1820-G5 radial engine of 1000 hp take-off power was selected as power plant. The Cyclone engine was rated at 850 hp at 6000 feet and 750 hp at 15,200 feet and featured a two-speed supercharger, Chandler Groves automatic mixture control and powered a three blade constant speed propeller. The engine was mounted on a welded chrome molybdenum steel tube structure, which was bolted to the fire wall. The engine could be replaced by detaching the four bolt attachment to the fuselage. To safe weight, the plane would not be equipped with a radio, and lacked additional armour or fuel tank protection. The tail wheel remained fixed.

This exploded view of the CW-21B variant provides an impression of details such as armament lay-out and the fitting of the fuel tanks in the wing center section.

Right:
When retracted, the landing gear legs were covered by aerodynamic doors. Only the wheels protruded, which should minimize damage in a wheels-up landing.

(collection N. Braas)

Front view of the Pratt & Whitney R-1820-G5 engine. It was a standard civilian engine but was reliable and power was sufficient. The front view reveals the greater width of the starboard blast tube, suggesting that a .50 machinegun may have been fitted. *(collection E. Hoogschagen)*

This photo shows that the roof of the sliding hood was strengthened with four lengthwise frames. *(collection T. Postma)*

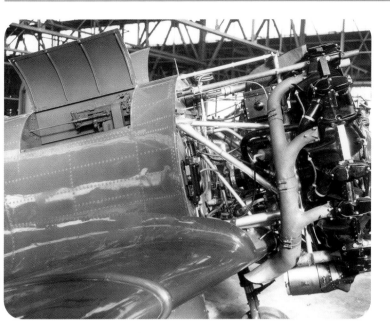

THE MILE A MINUTE FIGHTER

The CW-21 was advertised by Curtiss-Wright as the mile a minute fighter, referring to its unmatched climb rate. Although the climb rate of the CW-21 was indeed impressive, an actual climb rate of 5280 feet/minute was never officially recorded. A climb rate of 4800 feet/minute was published in 1940.

A CW-21B in an attractive yet fictional Royal Air Force livery, as illustrated by subcontractor Thompson products.

THE CW-21 'DEMON'

The CW-21 has been named Demon in many post-war publications. It has been commonly known like this for decades. But findings in Curtiss archive documents reveal that this is based on a misunderstanding of company data which was compiled in 1946. The CW-21 is mentioned in a listing of Curtiss-Wright designs. The root of the misunderstanding appears in a column with type names. The CW-21 is named Demon. (with a dot at the end). This was actually a reference to the purpose of the aircraft; it was a demonstrator. But a good fighting aircraft needs a name. The CW-21 was actually known as **Interceptor**, also in the Netherlands Indies. A much less appealing name compared to Demon… *

* As researched by Dan Hagedorn in the Museum of Flight archives, Seattle.

With the engine covers removed and armament bays opened, the armament of a single .30 and .50 machinegun is visible.
(collection M. Schep)

Bottom
The CW-21, piloted by Fausel.
(collection M. Schep)

SALES TO CHINA

After the test programme was completed, the CW-21 prototype was sent to China as demonstrator. It was transported to Rangoon (capitol city of Burma, nowadays called Myanmar) by ship and arrived there on 24 January 1939. From there the still crated aircraft was loaded on a barge and shipped to Bhamo. It was then transported by truck to Loiwing for assembly and was made available for demonstrations. Curtiss-Wright test pilot Robert Fausel flew the plane to Kunming on 2 March. The CW-21 was flown by Fausel in combat trials during the next day against French-built Dewoitine 510[*] and Soviet-built Polikarpov I-15[**] fighters.

Shortly after arrival in China military roundels and blue–white rudder stripes where painted on the aircraft. (collection N. Braas)

Chinese Nationalist air force markings were applied on the CW-21 prototype. The grey lower surfaces are an artists' impression. Profile art by Luca Canossa.

Performance of the CW-21 was found superior to the competing types. Although the Polikarpov I-15 proved to be highly manoeuvrable, it stalled earlier during prolonged manoeuvres. The CW-21 was able to exploit this and gain a firing position. A report on the demonstration suggests that the CW-21 demonstrator was not fully loaded and was only fitted with a pair of machineguns instead of four. Fausel also recalled no ammunition was loaded during the trials.

On 4 March Fausel flew to Chungking airfield, guided by a Douglas DC-2. Here, he demonstrated the CW-21 for the Chinese Minister of Finance and continued directly to Chengtu airfield. Here, the CW-21 and Curtiss 75Q were demonstrated on 16 March. Fausel flew a mock dog fight against a Polikarpov I-15bis fighter and demonstrated both Curtiss fighters for a second time on 22 March.

The CW-21 performed magnificently in the hands of Fausel and demonstrated to be a very manoeuvrable machine. It reached a very favourable maximum speed of 310 mph and an altitude of 10,000 feet was reached in two minutes 10 seconds. Main concern was that both the 75Q and CW-21

types may be branded as too complex for inexperienced Chinese pilots. Chinese pilots were allowed to fly both the Curtiss fighters after the 22 March demonstration. During this period, cracks started to appear in the Pyralin windscreen, which was not resistant to the Chinese climate. Fausel recommended to replace the windshield with a modified example made out of three pieces.

A DEMONSTRATOR KILL?

While a sale to China was negotiated, the single CW-21 was located at Chungking airfield, as was the Curtiss 75Q demonstrator. Chungking was frequently hidden in low clouds and thus relatively safe.

---

[*]   The Dewoitine 510 was a low wing all metal fighter with fixed landing gear, armed with two wing mounted machine guns and engine mounted 20 mm cannon. The open cockpit machine was first flown in 1932.

[**]   The Polikarpov I-15 was a biplane of mixed construction, fixed landing gear and armament of four 7,62 mm machineguns. Top speed 367 km/h (228 mph)

After returning to the United States Fausel was debriefed by a navy intelligence committee on 25 July 1939. In contrast to Fausels diary and letter correspondence, the report does not mention the flights, let alone the Japanese aircraft he shot down, which is strange. Such first-hand information would have been highly valuable. Fausel stated that he was allowed an inspection of a Japanese plane wreck, which was laying on a sand bar in a river. It remains unclear if the described interception actually took place, and how the inspected plane wreck connects to this interception.

Robert Fausel in the cockpit of the XP-55
Ascender fighter prototype.
(Missouri historical society collection)

 The Japanese did not appear over Chungking until May 3rd Fausel scrambled as quickly as possible after an air raid alarm and rushed to an altitude of 10,000 feet, where he intercepted a formation of Japanese bombers. He identified them as Italian built Savoia Marchetti SM.79 bombers but in reality, they must have been Fiat BR-20 bombers, which had been delivered to Japan. Fausel described making a head-on attack on one of the bombers. The CW-21 was armed with a pair of .50 machineguns which unfortunately jammed after the first pass. The bomber had been hit, though, and started to pour smoke from one of the engines and fell out of formation. Fausel followed the stricken machine down and saw the enemy plane belly land in a rice paddy.

The Curtiss 75Q demonstrator was lost in an accident on 5 May, killing the pilot in the incident. A period of difficult negotiations started. Finally, a contract was signed on 27 May for three CW-21s plus 27 to be assembled at the CAMCO facility in Loiwing. The Chinese demanded some modifications to the design. Among these

American GIs found the wreckage of a CW-21 in 1945. The foot step flap in the fuselage side reveals that this is the former NX19431

was the modification of the single piece windscreen into a three piece windscreen, an increased armament to two .30 and two .50 machineguns, and an option to carry two external fuel tanks. The demonstrator was handed over to the Chinese but it was quickly written off after a crash in June 1939.

The first of three production machines (c/n 21-2) was first flown by pilot Willis Wells on 20 March 1940, with civil registry NX19441. The second machine, c/n 21-3 and registry NX19442, followed shortly afterwards. It received fittings for carrying two external fuel tanks underneath the wing centre section to enlarge the fuel capacity. The third aircraft, c/n 21-4, was registered as NX19443 on 1 April.

These photos show that the production machines were generally similar to the first prototype. The windshield has been strength-ened and the sliding hood now has a coloured roof and only two horizontal frames.

Above: *(collection T. Postma)*
Middle: *(collection T. Postma)*
Bottom: *(collection E. Hoogschagen)*

Right page:
Above: (collection N. Braas)
Bottom: *The leading edge of the wing and engine cowl were painted aluminium dope.*
(collection E. Hoogschagen)

The three aircraft were to be armed with a standard armament of two .30 and two .50 machineguns, resulting in a loaded weight of 4250 pounds, an increase of 158 pounds compared to the prototype. Armament was not actually fitted however. The modified windscreens were not fitted either. Curtiss-Wright arranged for a demonstration before US Navy officials. C/n 21-3 flew to Anacosta Naval Air Station during the first days of April. Another demonstration at Wright Field was arranged, but on 5 April the plane was damaged during landing for a refuelling stop at Philadelphia airport. After repairs, all three aircraft were crated and shipped to Asia during mid-May.

The three aircraft arrived in Taungoo, 150 miles north from Rangoon during early 1940, and materials and components for assembly of the 27 CAMCO assembled planes reached Rangoon during the early months of 1940, until summer of 1940. But French and British attitude towards transport of armaments had changed since the contract signing, making transport to Loiwing impossible. The crated CW-21s and materials remained in storage for over a year.

Above:
NX19441, photographed from the rear cockpit of the CW-23 prototype.
(collection N. Braas)

Left page:
This front view shows that the landing gear struts were strengthened.
(collection N. Braas)

A close-up showing the two external fuel tanks fitted underneath c/n 21-3, NX19442.
(collection T. Postma)

Planned assembly of the CW-21 components was seriously delayed when a Japanese bombing raid hit the CAMCO facilities on 26 October 1940. By March 1941 no progress was made and it remains unclear if any aircraft were actually completed, although a report dated May 1942 suggests that some work had actually taken place. Two airframes were seen at Kunming airfield during summer of 1942, ready for test flights. It remains unclear if these machines have ever flown.

One of the disassembled aircraft in a ware house. The Chinese roundels have apparently been applied in the factory.

The third production CW-21 received civil registry NX19442. It could be fitted with external fuel tanks. Profile art by Luca Canossa.

A BRIEF FLYING CAREER

After being crated for some time, the CW-21s were finally purchased using Lend-Lease funds and drafted into the first American Volunteer Group (AVG), the legendary Flying Tigers, a fighter unit manned by US volunteers operating Curtiss P-40 fighters. It was commanded by Claire Chennault, who was personal advisor to Chang Kai-Shek. The AVG urgently needed a fast climbing fighter to deal with Japanese high altitude reconnaissance planes. The CW-21s were assembled and flown to the AVG training base Kyedew (Myanmar). After a short stay at Kyedew the three aircraft were attached to the AVG's 3rd squadron and were transferred to Mingaladon airfield north of Rangoon.

The planes were manned by pilots Erik Shilling, Kenneth Merritt and Lacy Mangleburg. After an uneventful stay, the detachment received new orders on 22 December. They were ordered to relocate to Kunming, China. This was the AVG's main operating base. The group flew to Kyedew the same day, and after a night's rest, proceeded to Lashio, roughly halfway between Rangoon and Kunming. During the flight to Lashio, the engine of Shilling's

A completely unmarked CW-21 on an airfield in Rangoon, prior to the fateful transfer flight to Kunming. (collection T. Postma)

The same plane as on the previous photo, seen together with an AVG P-40 being assembled

(collection T. Postma)

plane gave some issues. At Lashio, the group was advised to fuel the machines with 87 octane fuel instead of 100 octane fuel. After refuelling the three men quickly took off for the last leg of the flight to Kunming. Shilling again encountered engine difficulty, this time leading him to force land in rough terrain. After loitering over the area where Shilling landed, the two remaining pilots had to find a place to land. Their planes were running low on fuel, or possibly had engine trouble as well. The pilots may have been confronted with deteriorating weather. Merritt landed safely, but Mangleburg crashed into a mountain slope and was killed. All aircraft were destroyed in the incident.

DEVELOPMENT OF CW-22 AND CW-23

After completion of the CW-21 the St. Louis design team created two types of aircraft which are of note. First was the CW-22, an updated CW-19 design. The CW-22 featured the retractable landing gear of the CW-21 and was intended as advanced trainer and light reconnaissance plane. A converted CW-19 served as prototype. It could be equipped with a fixed, and one flexibly placed machine-gun in the rear cockpit and was able to carry a light bomb load. The CW-22 was sold to several foreign countries and was introduced into US Navy service as SNC-1 Falcon. The US Navy purchased 455 examples.

Another type on the drawing board was the CW-23, a combat trainer featuring a fully redesigned retractable landing

The Curtiss-Wright CW-22 was sold to a number of foreign nations and was purchased in substantial numbers by the US Navy as SNC-1 Falcon. (Missouri historical society collection)

This side view shows the streamlined fuselage lay-out of the CW-23 very well. Various refinements were used in the improved CW-21B. (US Air Force photo via M. Schep)

gear, which folded inward. This created a flush wing surface and eliminated the drag caused by the landing gear fairings. The fuselage featured a heavily modified cockpit section and a 600 hp Pratt & Whitney R-1340 Wasp engine. Company sources suggest it had a maximum speed of 325 mph (523 km/h). It was first flown in April 1939 with civil registry NX19427 and demonstrated to the US Navy, Canadian Air Force, and finally US Army Air Corps in May 1940, but none accepted it. Only one CW-23 was built. It was broken down in parts after the demonstration to the US Army Air Corps.

FURTHER DEVELOPMENT OF THE CW-21

Efforts to find foreign customers for the CW-21 were unsuccessful. The CW-21 was discussed with the Swiss air force technical commission late December 1938. The Swiss were originally interested in the Curtiss 75Q fighter but were impressed by the CW-21. Concern about the fact that it was still in development stage prevented further talks. Neither Curtiss types were

bought - the Swiss eventually selected the Messerschmitt Me 109E.

Despite this setback the CW-21 was developed further. A first proposal for a variant with a liquid cooled Allison engine was not worked out. Instead, several improvements found their way into a refined design, which was developed quickly

This photo reveals the flush finish of the CW-21B wing and the added tail fairing with semi-retractable tail wheel. (collection E. Hoogschagen)

after the CW-23. This type was known as CW-21B and utilised an inward retracting main landing gear, a tail wheel fairing with semi-retractable, steerable, tail wheel and hydraulically operated flaps which were all introduced on the CW-23. The fuel tanks in the outer wing panels were abandoned in favour of larger examples in the wing centre section.

Compared to the earlier CW-21, climb rate of the CW-21B was slightly influenced by the increased weight of the retractable undercarriage. The CW-21B achieved an initial climb rate of 4500 feet/minute and had an increased flying range.

Climb rate details on CW-21 are incompletely known. In comparison, the climb rate of the CW-21B was superior to contemporary fighters; Mitsubishi A6M2

	CW-21	CW-21B
Wingspan	35 ft (10,66 m)	35 ft (10,66 m)
Length	26 ft 4 in (8,03 m)	26 ft 4 in (8,03 m)
Height	8 ft 8 in (2,60 m)	8 ft 8 in (2,60 m)
Wing area	174.3 sq ft (16,20 m²)	174.3 sq ft (16,2 m²)
Weight – empty	3148 lb (kg)	3414 lb (1550 kg)
Weight – loaded	4180 lb (kg)	4500 lb (2042 kg)
Top speed		
- 12000 ft (3658 m)	304 mph (489 km/h)	312 mph (502 km/h)
- 17000 ft (5182 m)		315 mph (507) km/h)
Cruise speed	-	282 mph (450 km/h)
Initial Climb Rate		
-13.120 ft (4000 m)	-	4 min
-16.400 ft (5000 m)	-	5 min
Service ceiling	35.600 ft	34.300 ft (10455 m)
Range	530 miles	630 miles (800 km) at 282 mph (km/h)

Purchase of the Hawker Hurricane Mk I was contemplated but never materialised. On this photo the type is seen on display at the 1938 Paris air show. (collection E. Hoogschagen)

The Heinkel He 112 was another considered type. It was displayed in Holland and a number of Dutch pilots were allowed to fly the fighter. (collection E. Hoogschagen)

(16,400 ft in 5 min 55); Curtiss P-40E (15,000 ft in 5 min 10) Hawker Hurricane Mk II (15,000 ft in 5 min 54), but its top speed was lower, and at an lower altitude.

A DUTCH PURCHASE

Curtiss-Wright contract manager E.C. Walton attracted attention from the Dutch government in January 1940. At that time, the Netherlands was seeking reinforcements for its under-equipped Militaire Luchtvaart - M.L. for short (the Army Air Corps). A procurement programme was started as early as mid-1938 but little progress had been made. The Dutch

industry was unable to deliver a truly modern fighter at short notice. Possible purchase of foreign aircraft types such as Hawker Hurricane and Heinkel He 112 bogged down in discussion, lack of decisiveness and political unwill.

By January 1940 it had become evident that options to purchase a modern fighter type were becoming ever more slim. War had erupted in Europe on 1 September 1939, which had completely jammed any purchase option in an already overstrained market.

Walton was able to set up a meeting with general Petrus Best, commander of Dutch air defences. During this meeting the P-36

and P-40 designs were offered, and later the CW-21 was discussed, but Walton had urged Best to decide quickly. Walton said the offered CW-21s were already nearing completion and a decision should be made with the greatest speed. (In reality, the three CW-21s intended for China were indeed in final assembly stage, and production of parts was also well under way. After completion, the CW-21 production line would be idling.) To assure quick delivery, the aircraft should be accepted in as-is state. This, as Best recalled, sparked him to make a phone call directly to the Minister of Defence insisting on a quick handling of this opportune offer. It was decided to order 24 machines.

L'AQUILONE

26 LUGLIO 1942 - XX - SPEDIZIONE IN
ABBONAMENTO POSTALE - II GRUPPO
COSTA CENTESIMI 60

N° 30

Settimanale per i giovani

UNA "FOLGORE,, ATTACCA E ABBATTE UN "CURTISS WRIGHT 21 B,, *(disegno di M. Guerri)*

PERDITA DEL CONTROLLO

Sicuramente non diremo una cosa originale affermando che anche la flemma britannica bisognerà considerarla, nella storia di questa guerra, come uno dei tanti luoghi comuni miseramente sgonfiato. Quando si è favoriti dalla fortuna in maniera così scandalosa come lo sono stati gli inglesi per tre secoli, quando la tranquillità più sonnifera ti ha creato un senso della vita così facile, al di fuori di ogni preoccupazione quotidiana, non è detto che bisogna assolutamente essere inglesi per essere flemmatici. Basta salire su un tram e trovare un posto a sedere che subito ci meravigliamo del nervosismo di tutta quella piccola gente in piedi che si infastidisce per tanto poco.

Però, non c'è cosa peggiore per perdere la calma del non essere abituati ad essere nervosi. Alla prima occasione si diventa addirittura nevrastenici. E' quello che è capitato agli inglesi in questa guerra che ha segnato per essi il principio della vera scomodità. Guardiamo il caso delle navi, per esempio. Dopo essersi fatti venire i calli al cervello a furia di pensare che la Marina avrebbe deciso ancora una volta le sorti della guerra a favore di Sua Maestà britannica, è stato sufficiente che cinque o sei corazzate siano andate a fondo sotto i colpi delle bombe o dei siluri aerei, per cambiar parere con una frenesia tale da far impallidire noi stessi nervosissimi continentali e noi specialmente fantasiosi ed immaginosi italiani che sulla questione delle navi e degli aerei, con tutto quel po' po' che abbiamo combinato non vogliamo ancora affrettare azzardate

conclusioni. Diciamo dunque che anche la flemma, assai più meritata, farà parte dell'eredità che Albione lascerà ai popoli dell'ordine nuovo.

Ed ora, dopo le chiacchiere, i fatti alla mano:

Il giornale politico e laburista «News Chronicle» commentando la decisione americana di sacrificare le corazzate alle navi portaerei, così scrive:

«Il periodo delle navi da battaglia è finito. La potenza aerea deve prendere il posto della potenza navale.

Le battaglie del Mar dei Coralli e di Midway hanno indicato quali debbono essere le relazioni tra potenza navale e potenza aerea. Nell'azione del Mar dei Coralli le forze navali avversarie non si sono neanche scontrate: lo scontro si è sviluppato tra le opposte forze aeree.

E la voce del sangue ha fatto eco dall'opposta riva atlantica.

La Radio di Washington, commentando la notizia secondo la quale le autorità americane avrebbero deciso

di non costruire più corazzate, ma di rivolgere tutti gli sforzi nella produzione delle navi portaerei, così dichiara: «Questo significa che finalmente si sono comprese le lezioni impartite dagli avvenimenti del Pacifico. Si ha l'impressione ben precisa che le grandi corazzate siano vulnerabili e che esse si trovino sempre nel raggio d'azione degli aeroplani terrestri, anche quando navigano in alto mare.»

Insomma accadono dei fatti abbastanza sconcertanti. C'è la guerra

sui mari, avvengono tremende battaglie navali durante le quali le navi non sparano neanche un colpo e, tuttavia, vanno a fondo. Questo è accaduto nel Pacifico, questo è accaduto nel Mediterraneo durante uno dei numerosi scontri. E' dunque logico supporre che gli inglesi, che credevano di vincere la guerra perchè in possesso della più grande marina da guerra del mondo, abbiano perduto, assieme al controllo dei mari, il controllo di se stessi.

Discussions on the CW-21 purchase continued via the Dutch sales house Lindeteves-Stokvis. They offered a quotation on 1 April, which comprised a price per aircraft, without radio, oxygen supply, landing lights and armament, of $ 46,820. Curtiss-Wright would be able to deliver the mentioned equipment at an extra cost of $ 2,684. If the offer would be accepted, the first aircraft would be delivered 75 days after contract signing, and the last after 160 days. There were, however, no guarantees or penalty clauses and the factory would offer only the first production aircraft for acceptance trials – which would delay delivery by roughly 30 days. A down payment of 50% would have to be made.

A new quotation for the refined CW-21B variant was offered on 2 April, at a price of $ 53,320 per aircraft, which was agreed in oral form on 3 April with some modifications; the aircraft would need to be delivered with fittings for four FN Browning machineguns of 7.9 mm. These guns would be purchased separately at FN Browning in Belgium. The aircraft would need to be prepared for carrying two additional internal 16-gallon fuel tanks. Armour plating behind the pilot, oxygen supply, landing lights, radio equipment, a flare dispenser, fittings for a gun camera, propeller spinners and formation lights were included in the contract at extra costs. The contract included 6 engines, 6 radios, propellers and wings, engine parts, tools and 25% spares. The planes would need to be painted in camouflage paints and Dutch nationality markings. Delivery was scheduled between 15 July and 31 October. A first deposit was paid on 11 April.

The Fokker D.XXI fighter was the mainstay of the Dutch interceptor fighter force in 1940. It was flown valiantly but the German air superiority proved to be far too large.
(collection E. Hoogschagen)

A MODIFIED CONTRACT

The German attack on the Netherlands on 10 May 1940, and the Dutch surrender five days later, prevented deliveries. The Dutch government, which had gone in exile in the United Kingdom, made arrangements to transfer the CW-21B order to the Netherlands Indies Army for use by the Militaire Luchtvaart (M.L. for short - the Military Aviation Department of the Netherlands Indies Army). This was settled on 3 June. The 24 Curtiss-Wright CW-21B fighters were a most welcome addition to the inventory of the M.L.

A revised contract was signed on 11 July with some alterations. Most impor-

tantly, armament was changed to four fuselage mounted .303 in machineguns, the standard calibre used by the M.L. The actual armament was purchased separately. It is intriguing to mention that the CW-21B was not fitted with a gunsight in the cockpit. Curtiss-Wright advised to install a ring and bead visor, but this was ignored. Instead, a small pin sight was fitted, just aft of the upward folding engine cowlings. The pilot had to use the long engine cowl hinge, which ran on top of the nose, as aiming aide. Tracer ammunition would provide a means of adjusting the pilots' aim. The additional 16-gallon fuel tanks would not be fitted.

Left page:
An intriguing Italian propaganda drawing. It shows a CW-21B with temporary factory registry, the United States' roundel and Dutch camouflage scheme of brown, green and beige. It has been shot down by an Italian fighter!
(collection E. Hoogschagen)

The first production machine with temporary factory registry C-338. It is piloted by Robert Fausel, and is fitted with a tailwheel fairing, which was later deleted.
(collection T. Postma)

A relatively simple short range Western Electric VHF radio transmitter/receiver was installed, with dashboard fitted beneath the instrument board. The radio equipment itself was placed behind the pilot. Its frequency could be changed by exchanging radio crystals. This proved to be a time consuming process when in service. A wire antenna ran from the starboard fuselage side to the tail fin, and from there to the starboard wingtip.

The first two aircraft would be delivered as early as August, the next twelve during September and the final ten in October. The first machine – c/n 2852 – was delayed. It made its maiden flight in mid-September with temporary factory registry C-338. This, and the next CW-21B, c/n 2853 (temporary registry C-339), were used for performance trials which were flown by test pilot Baley. The aircraft would receive definitive registries CW-343 to CW-366.

An overview of the instrument panel with the radio dashboard fitted underneath. (collection T. Postma)

Bottom:
The second production machine photographed on the factory grounds on 10 October 1940. It is now fully painted and with definitive markings.
(collection T. Postma)

CW-363 was shown to the public when the new Curtiss-Wright facilities at Lambert Field were opened on 19 November 1940.

(Missouri historical society collection)

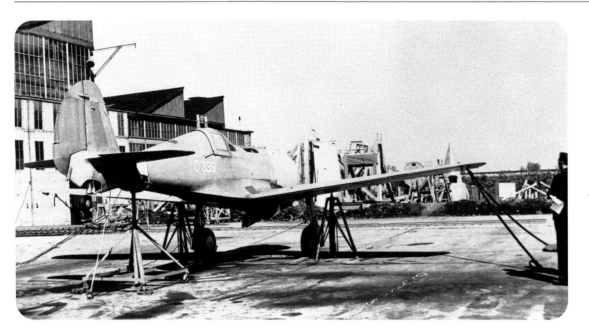

The still unpainted second production machine during gunnery trials on 8 October 1940. C-339 would be delivered as CW-344.
(collection T. Postma)

COLOURS AND MARKINGS

The aircraft would originally be painted according to Dutch LVA specifications for use in the Netherlands. This scheme may not yet have been determined, but would most likely comprise the standard three colour camouflage of brown, green and beige tones. The nationality markings, an orange triangle with black rim, would be placed on fuselage and upper and lower wing surfaces. Additionally, the tail rudder would be painted orange with black rim.

After the ordered aircraft had been taken over for delivery to the Netherlands Indies, a new paint scheme was agreed. This was based on work of the Centraal Bureau Camouflage (Central Camouflage Department). This department, subordinate to the Ministry of the Colonies, studied effective use of camouflage. Research included the camouflage of buildings, army equipment and aircraft. Another expertise was the planting of trees and bushes around the airfields in order to blend them in with the surroundings.

The M.L. had studied the camouflage for aircraft earlier. The results were incorporated in the work of the Centraal Bureau Camouflage. As early as 1938 a shade of green had been determined and a stock was already available. This colour was simply referred to as camouflagegroen (camouflage green). This was later supplemented by a darker tone of green (donk-

Assistant manager of the Curtiss–Wright factory William E. Nickey prepares for a test flight.
(collection E. Hoogschagen)

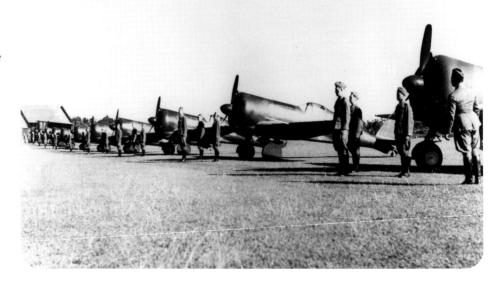

CW-21Bs lined up at Andir airfield, April 1941. The building in the rear is camouflaged according to the work of the Centraal Bureau Camouflage. Ground personnel have gathered in front of their assigned aircraft. (collection Nederlands Instituut voor Militaire Historie)

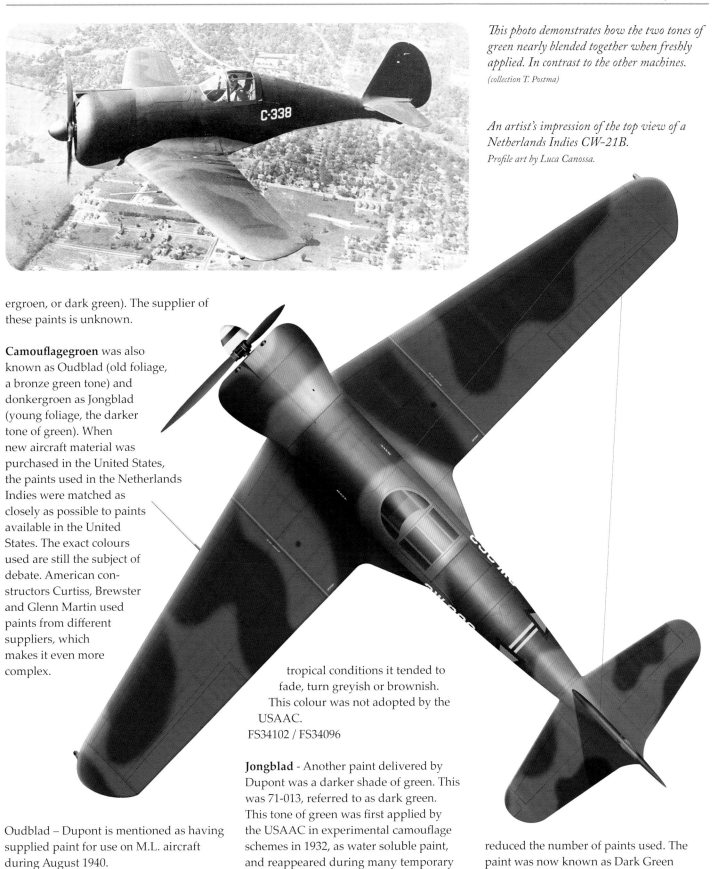

This photo demonstrates how the two tones of green nearly blended together when freshly applied. In contrast to the other machines.
(collection T. Postma)

An artist's impression of the top view of a Netherlands Indies CW-21B.
Profile art by Luca Canossa.

ergroen, or dark green). The supplier of these paints is unknown.

Camouflagegroen was also known as Oudblad (old foliage, a bronze green tone) and donkergroen as Jongblad (young foliage, the darker tone of green). When new aircraft material was purchased in the United States, the paints used in the Netherlands Indies were matched as closely as possible to paints available in the United States. The exact colours used are still the subject of debate. American constructors Curtiss, Brewster and Glenn Martin used paints from different suppliers, which makes it even more complex.

tropical conditions it tended to fade, turn greyish or brownish. This colour was not adopted by the USAAC.
FS34102 / FS34096

Jongblad - Another paint delivered by Dupont was a darker shade of green. This was 71-013, referred to as dark green. This tone of green was first applied by the USAAC in experimental camouflage schemes in 1932, as water soluble paint, and reappeared during many temporary paint schemes throughout the 1930s.

The paints used were standardised during 1938 and Air Corps Material Command

Oudblad – Dupont is mentioned as having supplied paint for use on M.L. aircraft during August 1940.
The paint which was matched to the Oudblad variant was 71-047 light green. When freshly painted this was a vibrant tone of green, but when exposed to

reduced the number of paints used. The paint was now known as Dark Green shade 30. Within the Curtiss company this colour was simply referred to as 'green'.
FS34092

THE M.L. — SHORT INTRODUCTION

An officer of the Royal Netherlands Indies Army voluntarily earned his flying license in 1911. The military staff was still reluctant towards aviation, but small numbers of aircraft were ordered for experimental use.

Right: The first large-scale aircraft purchases were made after World War One. Twenty-six De Havilland DH-9s were bought, the first of these entered service in 1919. They were supplemented with 10 locally built machines. Fokker D.VII, C.IV and C.V were ordered in the Netherlands. The Great Depression halted further purchases.
(collection Nederlands Instituut voor Militaire Historie)

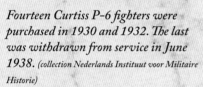

Fourteen Curtiss P-6 fighters were purchased in 1930 and 1932. The last was withdrawn from service in June 1938. (collection Nederlands Instituut voor Militaire Historie)

During the aftermath of the crisis of 1929 and under the Japanese threat a new defence doctrine was developed. The first line of defence of the vast archipelago of islands shifted from an expensive fleet of navy ships to a fleet of bombers. The Glenn Martin B-10 was found suitable and 120 aircraft were purchased in four production variants. The introduction of the Glenn Martin WH-139 was a major shift in defence policy. Purchase of a modern fighter type was neglected. In fact, there was only a single fighter in service from June 1938 until June 1940. But this Fokker D.XXI only flew sporadically for test purposes! (collection Nederlands Instituut voor Militaire Historie)

No fighters were acquired until a series of 20 Curtiss H-75A-7 Hawk fighters was bought in 1939. The first Curtiss H-75A-7 fighters arrived in June 1940. It was hoped that an extra batch of 28 Curtiss Hawk H-75A-4s, originally intended for France, could be purchased for equipping a full squadron, but this purchase was embargoed by the US government. 16 of the original 20 machines were still in service on 7 December 1941.

(collection E. Hoogschagen)

With the threat of war becoming ever more realistic, the need for modern aircraft resulted in an order for 92 Brewster Buffalo fighters in three variants. Of these, 26 could not be delivered to the Netherlands Indies anymore. 63 were still in service on 7 December 1941. Other selected types – Bell P-39, Curtiss P-40 and Hawker Hurricane – were unavailable.

(collection Nederlands Instituut voor Militaire Historie)

This factory photo clearly illustrates the jongblad coloured propeller and spinner, instead of the more common black. The tips remained unpainted. (collection T. Postma)

Center: *The orange triangle with a 10 cm thick black rim was the Dutch nationality marking since October 1939. The CW-21Bs were almost devoid of service markings. 'Opstap' was stencilled on both ends of the wing centre section, to indicate a walking area for maintenance crew. (collection Nederlands Instituut voor Militaire Historie)*

The lower surfaces of the wings and fuselage were kept bare aluminium. The propeller and spinner were painted Jongblad. To improve the camouflage of the aircraft the nationality markings were only placed on the lower wing surfaces and fuselage. The serials were applied on the fuselage in white, ranging from CW-343 to CW-366.

INTO SERVICE

The first CW-21Bs were delivered two months later than originally agreed; the first six were handed over during October, fourteen during November and four during December 1940. The aircraft were then shipped to the Netherlands Indies, the first machines arrived during November.

The machines were assembled at Andir airfield by the Technische Dienst (T.D., Technical Service) in the period of December 1940 up to February 1941. Curtiss-Wright technicians and test pilot W.C. Bryan were stationed in the Netherlands Indies to assist during test flying. After assembly the aircraft were assigned

to the 1e Proefvliegafdeeling (1st test unit), which was part of the 1e Jachtafdeeling (1st fighter unit) flying Curtiss Hawk. 23 aircraft were taken into service, a 24th was used as instructional airframe at the Monteursschool (mechanics school) at Andir.

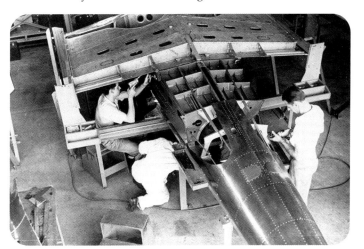

The fuselage and wing centre section were built as a single piece. Despite a strict contract, building and delivery of the 24 aircraft were delayed by several months.

(collection T. Postma)

Although still unarmed, the first four machines were flown alongside five Curtiss 75A-7 Hawks and a single Fokker D.XXI fighter.[*] During the first weeks of flying a first setback was suffered, when an Interceptor crashed at Andir after having engine trouble. The plane was heavily damaged. The fuselage had broken just in front of the tail and the area between cockpit and tail had to be extensively rebuilt, which took several months.

The first operational squadron to receive the Interceptor was the 2e Jachtafdeeling (2nd fighter unit), which was activated on 1 February. The unit was renamed as 2-Vl.G.IV (2e Afdeeling, IVe Vliegtuiggroep – 2nd squadron, IVth Aircraft Group) on 1 March. First Lieutenant A.A.M. van Rest became the unit's CO. Thirteen pilots were assigned to the new unit. Many had experience on the Hawk, and those who had to convert from the twin-engined Martin WH-139 bombers first trained on the Ryan STM trainers at Andir.

---

[*] A single Fokker D.XXI had been delivered for trials in March 1937 but failed to meet expectations. It was sporadically flown as advanced trainer and finally taken out of service in March 1941.

CW-357 being crated, 23 October 1940.

(collection T. Postma)

Center: *CW-354 had to be extensively repaired after a crash landing at Andir on 15 February 1941.*

(collection J. Terlouw)

The CW-21B was fitted with integral fuel tanks in the wing centre section, which allowed for maximum fuel capacity. These tanks were prone to leaking, with small quantities of fuel getting logged in the area between the wheel wells. These hidden fuel leaks were easily ignited when starting the engine. The construction of the wing did not allow for a modification of the fuel tanks, let alone replacing them with self-sealing examples. Shortly after entering service the fuel lines were insulated to prevent vapor lock, which could interfere with the fuel flow during engine start-ups. Issues with oil pressure were addressed as well.

Bottom: *A line-up of 23 aircraft at Andir airfield, 9 March 1941. A number of Hawks are just visible at the end of the line-up.*

(collection J. Grisnich)

Two Curtiss-Wright CW-22 Falcon advanced trainers were attached to 2-Vl.G.IV in April.
The two Falcons, registered CF-465 and CF-467, were used to prepare pilots for the transition to the CW-21B, and were used for armament trials when not flown by 2-Vl.G.IV pilots.

Aircraft mechanics were trained during the first months of operating, and by May 1941, 2-Vl.G.IV was declared operational. During the first half of May Sgt pilot G.M. Bruggink made a belly landing in his CW-21B after problems with oil pressure. He got distracted by a Ryan STM trainer which was in the landing circuit as well, and forgot to lower his landing gear on time, resulting in a pranged aircraft. The machine was repaired and, fitted with new propeller and engine, rejoined 2-Vl.G.IV in June.

Discussions on the purchase of 36 Curtiss-Wright CW-22 Falcons were well underway when the CW-21Bs were assigned to the M.L. The definitive CW-22 contract was signed on 4 June 1940. (collection E. Hoogschagen)

The available Interceptors were split between 2-Vl.G.IV and a new fighter unit, 1-Vl.G.V, which was formed on 1 June. The new squadron would transfer to the newly constructed airbase at Semplak. The move to Semplak commenced on 15 June. Some pilots remained at Andir to prepare for the delivery and initial training on the Brewster Buffalo. The first Buffalo fighters arrived in April 1941.

Both squadrons were assigned eleven aircraft – nine operational and two in reserve. 2-Vl.G.IV would transfer to Maospati airbase, which had opened on 28th March. The two CW-22s were handed over to 2-Vl.G.V when the transfer was announced.

The aircraft of 2-Vl.G.IV lined up at Andir, June 1941. The assigned mechanics and pilots have gathered before departure for Semplak airfield. This photo is part of a series covering this event. By this time, photographing had become very much restricted. (collection G. Tornij)

With engine covers opened, the Wright Cyclone R–1820 engine was easily accessible. This engine type was standard power plant for all first line fighters, bombers and transport aircraft.
The base of the propeller blades on the Interceptors remained unpainted.

(van Cornewald via M. Schep)

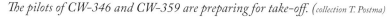

The pilots of CW-346 and CW-359 are preparing for take-off. (collection T. Postma)

1-Vl.G.V

The month of June was used to fit armament. A small number of machines rotated to Andir to have the guns installed in the nose of the aircraft. After returning to Semplak the synchronisation was calibrated and the guns were harmonised. 1-Vl.G.V pilots then took part in gunnery training at the auxiliary field of Anjer Lor. When gunnery practice commenced, pilots were confronted with exhaust fumes entering the cockpit, which forced pilots to fly with a partially opened cockpit hood. The cockpit ventilation was modified, solving the issue.

Ensign D.C.J. Joustra entering CW-363 at Semplak airfield. (Valk, via M. Schep)

CW-21B operations with 1-Vl.G.V proved to be short-lived. The new Brewster Buffalo fighters were introduced during July. Five CW-21Bs were flown to Maospati on 2 July, but one machine crashed into the ground during a low altitude pass over the field. The machine was completely destroyed and the pilot, ensign D.C.J. Joustra, was killed. The remaining four aircraft were handed over to 2-Vl.G.IV.

Six CW-21Bs remained with 1-Vl.G.V and were flown alongside the Buffalo fighters until September 1941. 1-Vl.G.V would act as test unit for the new Brewsters.

Fairly quickly after entering service, several airframes developed structural issues. During inspection skin buckling had been found in the tail and several

Members of ground personnel in front of CW-343, Semplak airfield, July 1941. At left mechanic J. Schellekens. (Schellekens, via M. Schep)

rivets had become loose. One airframe showed warping in the tail area. Curtiss-Wright was consulted and a modification, including reinforcement of the tail, was developed. Necessary materials were acquired in the USA during July and August. The last six aircraft of 1-Vl.G.V were the first group sent to Maospati for repairs and modification work. The other aircraft followed in groups until the last aircraft was modified in November.

Ensign R.A. Rothcrans in front of his machine. Rothcrans flew Brewster Buffalos during the fighting and was reported MIA on 24 January. (Patist, via M. Schep)

The six CW-21Bs which remained at Andir, with a line-up of Brewster Buffalos in the background, Andir airfield June 1941. (collection T. Postma)

2-VL.G.IV

Similar to 1-Vl.G.V the aircraft of 2-Vl.G.IV were fitted with armament at Andir. The pilots subsequently took part in gunnery training at Semarang airfield and gained first experience with gun camera firing. Training continued with shooting at a target drogue, towed by a Martin bomber. Ten new pilots were welcomed during July 1941 and a busy training period started. Several CW-21Bs attended the Army and Aviation days at Medan airfield (Sumatra) on 9 and 10 August. It was the only time the CW-21Bs ventured outside of Java. Two accidents occurred during the period of training

This photo reveals some interesting details, such as the landing lights, which folded outwards when activated. Magazine chutes are seen in the fuselage, just behind the landing gear doors. Flying in patrols of three aircraft was common. On 27 July an order was given to fly in a tactical formation of two aircraft. This was based on experiences during the air battles in Europe. (collection T. Postma)

CW-366 crashed during a training flight on 30 August 1941. (collection J. Terlouw)

A training flight was made on 30 August to practise for an air parade in honour of the 61st birthday of HM Queen Wilhelmina. Sgt Blans had to make a belly landing in CW-366 after suffering engine trouble. The aircraft was heavily damaged and was written off. Another incident occurred on 9 October. Sgt Janssen was killed when his plane, CW-361, entered an uncontrollable spin. This machine was written off as well. 2-Vl.G.IV now counted 16 CW-21Bs.

More structural issues were found late October, when cracks were discovered in the main landing gear trunnions of several CW-21Bs. A repair programme was started in November. The aircraft of 2-Vl.G.IV were sent to the T.D. workshop at Maospati in three groups. The six 1-Vl.G.V machines were repaired after their tail modification was completed. When these were completed, the aircraft were taken into service at 2-Vl.G.IV. The programme was finished in February 1942.

Nine CW-21Bs are joined by Brewster Buffalos on 1 September 1941 during the air parade in honour of HM Queen Wilhelmina.
(collection T. Postma)

2-Vl.G.IV was split up in three patrols, of initially four aircraft each. To distinguish these patrols the aircraft received a coloured fuselage band at the location of the orange triangle. This was done after the unit was transferred to Maospati in August 1941. Aircraft were flown by regular pilots and their name was painted on the nose of the aircraft in white. Many pilots rotated to other units, however.

First patrol;
White identification colour. A white band on the fuselage, combined with a white propeller spinner. Known registries are CW-357 (pilot Hermans), CW-358 (pilot v. Bers)

Second patrol;
Aircraft may not have carried identification colour.

Third patrol;
Yellow identification colour. Two thin yellow bands divided by a band of Jongblad (Young foliage green) of similar width. A yellow propeller spinner with Jongblad band in the middle. Known registries are CW-362 (pilot Haye), CW-363, CW-366 (pilot Blans). *
The patrol colours were used in radio communication too, with radio call signs referring to the patrol colours; red one, red two etc.

* The application of patrol bands was discussed by researcher P.C. Boer during interviews with former pilots and technicians during the 1980s and 1990s. Pilots recalled that the third patrol colour was actually azure blue, a paint which was mixed locally and was known as 'bleu' among the men.

CW-344 featured footsteps in the port side of the fuselage. The pilot is wearing a white flying cap. Pilots of some fighter units had personalised flight caps made in their patrol colours red, white or blue. (Patist via M. Schep)

CW-357 with the white patrol band and pilots name Hermans on the nose, photographed during gunnery exercises at Semarang, September 1941
(van Cornewald via M. Schep).

Right: *CW-362 with the yellow patrol bands on the fuselage. The white spinner of CW-358 in the foreground.* (van Cornewald via M. Schep)

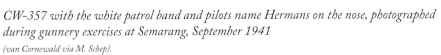

CW-362 of the third patrol with yellow identification colour. Profile art by Luca Canossa.

The patrol bands were removed between 5 February and 12 February. The pilots' names were retained. A hornet was painted on a single aircraft of 2-Vl.G.IV as squadron badge. Unfortunately no photographs of the Interceptor with this badge have surfaced.

DARK CLOUDS GATHER

Tensions were rising in Asia and the Pacific. The expansionist Japanese army high command gained much influence on the Japanese government. Japanese forces invaded China in 1937 and successfully held large portions of the country. Fears of further aggression were realistic.

The threat of war revealed the neglected state of the defences of the vast Netherlands Indies island archipelago. The M.L., despite the expansions which started during the mid-thirties, was still vastly underequipped.

Preparations for mobilisation of the Netherlands Indies armed forces started on 30 November 1941. The M.L. was mobilised on 2 December, followed by the Netherlands Indies Army on 12 December.

CW-358 seen during gun harmonisation on Semarang, September 1941. Note the white line on the rear of the propeller blade.
(van Cornewald via M. Schep)

A series of photos showing aircraft being readied for a mass departure. Some of the photos were released as press photographs on 12 August 1941.

(collection T. Postma)

Bottom: *(Collection J. Grisnich)*

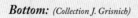

Worn off paint is noticeable on the footstep flaps in the fuselage sides of CW-356.
(Collection J. Terlouw)

THE CW-21B IN COMBAT

A relaxed pre-war photo of pilots gathered in front of their machines. (collection E. Hoogschagen)

Although the Japanese attack on Pearl Harbor is best remembered, this was part of a multipronged attack on targets throughout Southeast Asia and the Pacific. Allied defences were surprised and out-numbered on land, at sea and in the air. During a matter of weeks the defence of major strongholds collapsed. The Japanese forces seemed unstoppable…

During peacetime the main body of the M.L. was located on Java, with groups of aircraft venturing to other islands during exercises. When hostilities broke out, the M.L. was scattered throughout the archipelago, with detachments located on Ambon and Borneo. Several units flying Martin bombers and Buffalo fighters were sent to aid in the defence of Singapore. Similarly, some Australian units were located on Ambon and Timor islands. 2-Vl.G.IV remained stationed at Java, due to the CW-21Bs endurance restrictions. The

registries of aircraft involved in combat action or those which were lost during the fighting are not exactly known. Reports or pilot log books did not survive the war, and those that did do not provide full insight.

Twenty CW-21Bs were still in service, thirteen were assigned to 2-Vl.G.IV and seven were at the T.D workshop at Maospati. 2-Vl.G.IV was split up in three patrols of four aircraft. The first two patrols, I-2-Vl.G.IV and II-2-Vl.G.IV, relocated to Andir airfield, near Bandung. The third patrol, III-2-Vl.G.IV, moved to Tandjong Perak airfield, located near Surabaya.

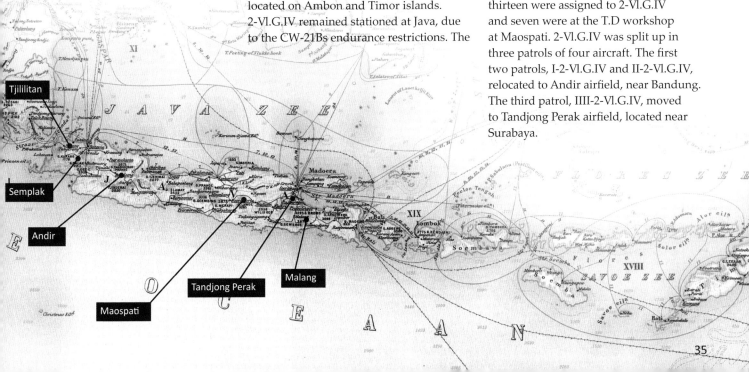

During the next weeks the patrols flew their regular exercises and some patrol flights along assigned parts of the coast. During this period work on the main landing gear trunnions continued, which meant that aircraft had to be ferried to and from Maospati airfield.

With the CW-21Bs remaining at Java, the first three weeks of war passed without much action for 2-Vl.G.IV. During January the M.L. was tasked to protect British shipping en route to Singapore. The first convoy passed through Dutch Indies waters late December and navigated through the Sunda and Banka Straits* on 1 and 2 January. The CW-21B patrols were standing by on readiness alert, while other units provided convoy patrols. II-2-Vl.G.IV moved to Palembang airfield, southern Sumatra on 2 January and remained on standby while the convoy progressed further north. A second convoy was covered on 4 and 5 January. When the third convoy passed through a scramble alert was sounded on 12 January and four Interceptors took off from Palembang airfield to search for a reported formation of Japanese bombers. Nothing was found.

The patrol returned to Tjililitan airfield (near Batavia) on 15 January, where the aircraft would continue their standby alerts when three more convoys steamed towards Singapore. By the end of January the entire 2-Vl.G.IV was relocated to Tandjong Perak airfield. Reinforcements, in the form of British, Australian and American aircraft, arrived on Java during the fighting. Five British radar posts were relocated to Java**, which gave the defenders a little time to respond.

---

* Sunda Strait is the sea strait between Java and Sumatra. Banka Strait is located further north between Sumatra and the island of Banka

** Sources are contradictory on the number which could be installed.

A recently assembled CW-21B photographed at Andir airfield. The CW-21B was the fastest climbing fighter in M.L. inventory, but not the fastest in level flight. Despite its excellent climb rate and manoeuvrability the pilots were seldom able to exploit the qualities of the fighter in combat. The pilots were instructed to avoid a direct confrontation with enemy fighter aircraft.

(collection Nederlands Instituut voor Militaire Historie)

A nice shot of CW-356. Note the warped fuselage skin just behind the engine cover. Skin warping and structural issues were also encountered in the tail, necessitating a repair and strengthening programme. (collection E. Hoogschagen)

subishi A6M fighters, took off from bases in Borneo and Celebes for an attack on Malang airfield, the Surabaya naval docks area and Madura Island. After being alarmed, twelve CW-21Bs from 2-Vl.G.IV and seven Hawks from 1-Vl.G.IV were scrambled. They were joined by six US P-40s of the 17th P.S. By the time the pilots managed to make contact with the enemy aircraft, these had already released their bombs and were returning home. The Interceptors split up in three sections. Each got involved in a dogfight with a large group of A6Ms. The fighting took place over Madura Island.

1st Lieutenant Anemaet, squadron CO, led the 1st patrol of four Interceptors, but none

of the pilots was able to make contact with enemy aircraft. They finally got involved in the fight when they came in to land at Perak airfield. Sergeant van Balen was shot down and killed, but sergeant Beerling broke off his landing and fired all his ammunition in a dogfight with Zeros. Out of ammunition, Beerling landed at Ngoro airfield. 1st Lt. Anemaet did not make contact with enemy aircraft and landed his Interceptor at Perak airfield, which had been bombed during the dogfight. Anemaet's plane rolled into a bomb crater after landing, putting the plane on its nose. Anemaet was uninjured, but his fighter was heavily damaged. Sergeant Dejalle landed at Perak without trouble.

Japanese Mitsubishi C5M Karigane*
reconnaissance planes appeared over eastern Java on 1 February with the goal to determine number of retreated US airplanes at Malang airfield. 600 kilometres further east, Maospati airfield was reconnoitred as well. CW-21Bs were scrambled but none of the pilots was able to intercept the intruders. Maospati airfield was attacked twice the next day.

On 3 February a mixed force of Mitsubishi G3M Rikko bombers, escorted by Mit-

*1st Lieutenant R.A.D. Anemaet was squadron leader and patrol commander. He made two combat flights. Afterwards he was involved in acceptance flights of Hawker Hurricane and Curtiss P-40E. He continued to fly P-51D Mustangs after the war.
(collection Nederlands Instituut voor Militaire Historie)*

*   The C5M would receive the code name Babs during the war. The Mitsubishi G3M Rikko would become known as Nell and Mitsubishi A6M as Zeke, but better known as Zero

Sergeant F. van Balen was shot down and killed on 3 February 1941. (www.dornierdo24k.nl, via E.W. van Willigenburg)

Sergeant F. Beerling partici-pated in the dogfights of 3 and 5 February. He was then assigned to T.D. Andir and test flew repaired aircraft. He also was involved in testing the newly arrived Hurricane fighters. He survived the war and flew P-51Ds up to 1950. (collection Neder-lands Instituut voor Militaire Historie)

Sergeant N. Dejalle in front of a Koolhoven F.K. 51 trainer aircraft. Dejalle made a single CW-21B combat flight on 3 February and another single sortie in a Hurricane. He was tasked with test flying P-40Es. He survived the war as POW and, like other squadron mates, continued to fly P-51Ds after the war. (Dejalle via M. Schep)

HERMANS   CW-357

Ensign A.W. Hamming had flown Buffalo and Hawk fighters before joining 2-Vl.G.IV. Hamming was involved in the combats on 3 and 5 February and flew a single Hurricane sortie. He was made POW and was liberated in Singapore. Hamming became the Nether-lands' first crop duster pilot, but was killed in a flying accident on 5 July 1950.

(collection Nederlands Instituut voor Militaire Historie)

The 2nd patrol was attacked from above and behind. Reserve 2nd Lieutenant Kingma, leader of the group of four fighters, made a sharp evasive turn, which caused the patrol to break up. Kingma scored hits in the tail of a Zero, while his wingman, sergeant Haye in CW-360, claimed a Zero as shot down, it was seen going down in flames. Kingma spotted a second group of what he thought were Zero fighters and could attack one of the machines from short distance. He claimed this aircraft, possibly a C5M, as shot down. Kingma was shot down right after his success by a fighter he had not noticed. Badly burnt, he managed to bail out of his plane. Haye similarly spotted more enemy fighters and while lining up for an attack on two machines, he was fired at by a third, which he had not seen. With oil and hydraulic lines shot through, he had to dive to safety. After jettisoning his cockpit

hood in order to bail out, Haye realised his plane was still responsive and made a successful landing at Ngoro airfield. There was no ammunition available at Ngoro, so Haye handed over his remaining rounds to Beerling, who had landed on the field too. The other pilots, ensign Hogenes and sergeant Halberstadt, were shot down and killed after they got separated from the other planes.

First Lieutenant Bedet was leading the 3rd patrol, a section of three Interceptors with sergeant Roumimper and ensign Brouwer, when they were attacked by nine Zero fighters. Brouwer, who was at the rear of the formation and acted as lookout, could warn Roumimper before one of his wing tanks was shot through. He was able to empty this tank and exploited the climb qualities of his plane. He placed himself behind two enemy fighters, shooting one

Center: Ensign J. Hogenes was shot down and killed together with his wingman Halberstadt.

Right: Sergeant H.M. Haye was in action against Zero fighters on 3 February and Ki–43s on 24 February. He made three more flights during the last week of fighting. (collectie Nederlands Instituut voor Militaire Historie)

Left: 2nd Lt. J. Kingma was formerly trained as Glenn Martin pilot and retrained as fighter pilot on the Curtiss Hawk. He made a single combat flight. He was shot down after claiming an enemy aircraft shot down and was made POW after the capitulation of the Netherlands Indies.
(collection Nederlands Instituut voor Militaire Historie)

of them down. His own plane was shot up during the dogfight and Brouwer was wounded in a leg. He had to belly land on Madura Island.

Roumimper got entangled in a frantic dogfight. His aircraft was hit and parts of the windscreen were shattered, forcing him to disengage and land at Maospati. As soon as he was able, he jumped out of his plane. Pursuing Japanese fighters again attacked his Interceptor, completely wrecking it on the ground. Bedet, flying a CW-354, did not fare any better. His plane was shot up in a dogfight. A 20 mm shell hit the windshield, wounding Bedet in his arm and chest. Bedet belly landed at Perak airfield and was barely pulled out of the wreckage and to safety by airfield personnel before his plane was strafed on the ground as well.

Sergeant R.C. Halberstadt, seated, second from left, was 23 years old and just married for three months when he was killed in action.
(www.dornierdo24k.nl, via E.W. van Willigenburg)

A pilot climbing aboard CW-350.
(collection E. Hoogschagen)

A pilot prepares for take-off. Note the weathered paint on the leading edges of the wings.
(collection E. Hoogschagen)

Ensign Hamming was on leave on 3 February but returned to the airfield as soon as he heard the air raid sirens. He took off in a reserve plane. He was unable to catch up with the rest of the patrol, and found himself amidst a formation of Zero fighters. After firing a short burst at one of them, he dived to safety and landed at Maospati airfield.

Seven out of thirteen CW-21Bs were lost on 3 February, and another was damaged beyond repair and given up a few days later. Roumimper's plane was sent to the T.D. workshops at Maospati, where it was cannibalised for parts.

2-Vl.G.IV lost three pilots, two were badly injured and a third had light injuries. The pilots claimed two Zeros and a C5M shot down and a third Zero damaged. The Japanese bombers were able to reach their designated targets without serious hinder. All targets were bombed with success, while escorting fighters dealt with the intercepting Allied fighters. Numerous American and Dutch aircraft were lost on the ground or at the seaplane harbour of Surabaya.

The fourth member of the 3rd patrol, ensign Dekker, was delayed but managed to get airborne in a CW-347. He joined a group of aircraft which he identified as friendlies but got mixed up in a formation of Zero fighters. After a lucky shot at one of these, Dekker's plane took a hit in the oil tank which caused a spray of hot oil. Dekker tried to shake off his attackers by escaping into a cloud. With his engine seized, he managed to escape with a near vertical dive and belly landed at a beach of Madura Island.

Ensign J. Brouwer was a former Brewster Buffalo pilot and made a single combat flight on 3 February. He was wounded by shell fragments and suffered a concussion during a crash landing on Madura Island.
(collection Nederlands Instituut voor Militaire Historie)

Ensign D. Dekker was killed in action on 21 February. (collection Nederlands Instituut voor Militaire Historie)

Sergeant O.B. Roumimper completed six combat missions in the CW-21B. He survived the war as POW and rejoined the army in 1946 as pilot on North American P-51D Mustang.
(collection Nederlands Instituut voor Militaire Historie)

1st Lt. W.A. Bedet flew one combat mission on 3 February, during which he was wounded. He was made POW and continued to serve in the Royal Dutch Air Force after the war. (D. Bakker via M. Schep)

Six Interceptors were available again on 5 February, three of these had just returned from the T.D. workshops at Maospati. Five were combat ready and scrambled during the early morning as a new wave of Japanese aircraft approached Surabaya. The CW-21Bs, led by 1st Lieutenant Anemaet, were accompanied by two Curtiss Hawks and seven P-40Es.

One of the CW-21B pilots, ensign van der Vossen flying CW-345, had to retreat from combat. The engine of his plane did not run smoothly and there were problems with the armament. Van der Vossen made a safe landing at Kediri airstrip but his plane was damaged and had to abandoned. The four remaining pilots were confronted by a formation of an estimated 26 Zero fighters. First Lieutenant Anemaet managed to get into a firing position after an Immelmann manoeuvre and successfully fired at the leader of a formation, which he claimed as destroyed.[*] The two other Japanese flyers immediately responded, and Anemaet's plane was heavily damaged in the following combat. The tail of his machine was severely hit, a wing and the engine area

CW-356 with the Merapi volcano, near Semarang, in the background. (collection E. Hoogschagen)

Despite the difficult circumstances, T.D. personnel was able to keep as many aircraft in the air as possible. Shortage of spare engines and parts was however a growing strain. (movie still, collection Nederlands Instituut voor Militaire Historie)

Ensign van der Vossen during his pilot training, in a Koolhoven F.K. 52 trainer. (Patist, via M. Schep)

---

* Anemaet claimed a fighter, but he very likely damaged a C5M reconnaissance plane, which broke off after receiving hits

were damaged as well. Anemaet managed to return to Perak airfield and landed without incident. The other pilots, sergeant Beerling, sergeant Roumimper and ensign Hamming, dogfought a formation of six Zero fighters. There weren't any losses on either side and the three pilots all returned to Perak safely. Two of the planes were damaged, sergeant Beerling returned with an undamaged plane.

After this second combat 2-Vl.G.IV was able to repair its aircraft during the next days. Six aircraft were available again on 7 February, mainly due to the return of three aircraft from T.D. workshops after modification work on the main landing gears. Two machines were undergoing repairs at T.D. Maospati. These may have been CW-343 and CW-345.

CW-21B CW-358 with pilot's name v. Bers on the fuselage. Van Bers was reassigned to fly Brewster Buffalo fighters, and went missing on 4 December 1941.
Profile art by Luca Canossa.

Although a formation of four Japanese aircraft made a reconnaissance flight over Maospati airfield on 7 February, no orders to intercept them were given. Four aircraft of 2-Vl.G.IV scrambled during the morning of 8 February, together with four P-36 Hawks and US P-40Es from Ngoro. They tried to intercept a reported group of Japanese aircraft approaching from the north, but no intruders were found. Only afterwards it was learnt that a group of US B-17 bombers was possibly misidentified as enemy aircraft.

A rather relaxed pre-war photo of Lieutenant Boxman
(collection Nederlands Instituut voor Militaire Historie)

A very rare colour photo of a CW-21B, photographed at Andir airfield. The orange toned hood top is clearly noticeable. (via J. Grisnich)

Malang airfield was attacked on 9 February. Available CW-21Bs were ordered to patrol over Surabaya naval base and wait for Japanese aircraft returning from their attack run. No enemy aircraft were sighted, and the Dutch fighters returned to base.

Japanese reconnaissance aircraft were active on 10 and 11 February as well, reconnoitring Surabaya, Maospati and Malang. Despite being scrambled, the pilots were unable to make contact with the enemy aircraft. All six remaining aircraft were relocated to Maospati, and transferred to Andir airfield the next day.

The Japanese landed on Sumatra on 14 February. The airfield of Palembang was now available to the enemy, which enabled both bomber and fighter aircraft to reach the bases of Andir and Kalidjati in the more eastern part of Java.

Another transfer was made on the 15th. This time, all available CW-21Bs were formed into a patrol operating from the Boeabataweg, an improvised landing strip on a road near the city of Bandung. Lieutenant W. Boxman took command of the patrol. Five aircraft were in service, with one in reserve. Another was at the T.D. workshop at Maospati

Japanese Mitsubishi Ki-46 Shitei (code named Dinah) high speed reconnaissance planes appeared over western Java on 18 February. The intercepting Buffalo fighters could not reach the altitudes at which the Japanese were flying. Although the RAF Hurricanes and CW-21Bs were able to reach these altitudes, the Dinahs were too fast and all attempts to intercept the intruders failed.

Aircraft were scrambled several times on 19 February, but each time without results. At 15h30, three Interceptors scrambled after a fourth alarm. Twelve Brewster Buffalo fighters of 1-Vl.G.V had taken off earlier to intercept a large formation of aircraft. The fighters were directed towards a reported formation of Japanese bombers, but the CW-21B pilots were ordered to set a different course before they were able to make contact with the enemy. They were now ordered to return

to Andir, to intercept Japanese bombers. The reported Japanese bombers, nine Kawasaki Ki-48 Sokei (code named Lily), were able to reach Andir airfield before the Interceptors, causing extensive damage. The Buffalo formation in the meantime entered a dogfight with Nakajima Ki-43 Hayabusa (code named Oscar) fighters. Three Buffalos were lost, and one Ki-43 was claimed shot down.

The engine of CW-363 malfunctioned and was sent to the T.D. workshops at Andir for repair. CW-364 was already at the T.D.

Work at the T.D. workshops at Andir continued until the very last days of combat.
(movie still, collection Nederlands Instituut voor Militaire Historie)

workshops at Andir to have a landing gear issue solved. After a check a test flight was made, but a wheels up landing had to be made. The plane was dragged off the field by a tractor, loaded on a train and transported to Maospati. Ensign van der Vossen flew one CW-21B to Maospati for repair work on the engine. Van der Vossen made the return flight to the Boeabataweg strip in an Interceptor which was held in reserve.

Kalidjati airfield was attacked by a Japanese formation of ten Ki-48 bombers and escorting fighters on 20 February.

Although a Buffalo patrol was scrambled, followed by a patrol of Interceptors, none were actually ordered to engage the attackers.

During the late morning of 21 February a formation of Japanese bombers, and its escort, was reported above Batavia, heading towards the east. Seven Buffalo fighters scrambled, followed by three CW-21Bs. A fourth Interceptor remained on the ground - there was no pilot available.

One of the CW-21B pilots, 2nd Lieutenant Dekker, got separated from his squadron mates during the climb to combat altitude and decided to return to Andir airfield. The lone plane was caught by Japanese fighters and shot down, killing Dekker. The two remaining Interceptors entered combat with Ki-43s, but without results or losses. One CW-21B, just reported ready after maintenance, was destroyed on the ground during an air raid at Maospati.

The Interceptor patrol, down to four aircraft, was transferred to Andir airfield during the evening of 22 February.

The Boeabataweg strip would now be exclusively used by transport aircraft. The Interceptors were placed in aircraft pens in the southeastern corner of Andir airfield. Three CW-21B, flown by ensign van der Vossen, sergeant Haye and sergeant Roumimper, scrambled after an air raid alarm during the early morning of 24 February. They were joined by four Buffalo fighters. Due to poor weather the alarm had sounded late. The fighters were immediately caught in a dogfight while still climbing for altitude. The fight ended without results. While the dogfight developed, 17 Ki-48 bombers managed to bomb Andir unnoticed by the intercepting Dutch fighters. Air control sent instructions to the three CW-21B pilots to proceed to Batavia, but Haye had to land at Andir low on fuel. Van der Vossen and Roumimper proceeded on their given course. Unfortunately the two planes were misidentified by an RAF Hurricane pilot. Van der Vossen's plane was fired at and damaged. He made an emergency landing at Kemajoran airfield. Roumimper followed his lead and joined him at Kemajoran. Van der Vossen's machine was transported to Maospati by train.

RE-EQUIPMENT WITH HURRICANE FIGHTERS

2-Vl.G.IV received a dozen Hawker Hurricane Mk IIB fighters as replacement for the lost CW-21Bs. The Hurricanes were handed over on 15 February. Converting to the British fighters was challenging; the M.L. was completely orientated on American aircraft. Problems with radio equipment had to be solved and supply of 100 octane fuel, glycol cooling liquid, tools and spares had to be arranged. Some 40 men of RAF ground personnel were made available for maintenance assistance. To gain manoeuvrability and a higher top speed, the Vokes sand filters (the filter housing was retained) and the four outer wing mounted machineguns were removed. The squadron was led by 1st Lieutenant Anemaet and pilots from 1-Vl.G.IV and 2-Vl.G.IV were selected to fly the Hurricanes. Bruinink and de Wilde, pilots from 2-Vl.G.IV, joined them.

Two aircraft were unfortunately involved in accidents during conversion training. Eight machines made the first operational patrol flight on 25 February. They were attacked by Ki-43 fighters when the formation came in to land at Kalidjati. Two machines were heavily damaged in forced landings. These were damaged further on the ground during a second attack on the field and had to be written off. The remaining aircraft made a transfer flight to Ngoro field the next day. Seven Hurricanes were in action on 1 March. They made strafing attacks on Japanese landing barges and troops. All but three aircraft were damaged, one made a forced landing in a rice paddy. Two aircraft landed at Maospati, the others returned to Ngoro. The aircraft were readied for a new mission, but Ngoro was attacked by two Zeros during refuelling. Two Hurricanes remained undamaged and were immediately withdrawn to Maospati. None of the surviving aircraft made another operational flight. At least one aircraft was captured intact by Japanese forces.

To avoid unwanted Japanese attention, the twelve Hurricanes which were allotted to the M.L. were assembled on the roadside.
(collection Nederlands Instituut voor Militaire Historie)

On 26 July 1941 M.L. pilots had been ordered to fly in pairs, based on experiences in the European air war. Despite the tactical change, the inexperienced pilots were no match for the highly disciplined and combat hardened Japanese pilots. (collection T. Postma)

A DESPERATE WEEK

During the fighting some incidents led to a reconsideration of the orange triangle as nationality marking. The orange triangles were too similar to the bright red roundels carried by Japanese aircraft. This led to various incidents of friendly fire. On 24 February it was decided that the orange triangles were to be replaced by a new marking; the Dutch flag in national colours red, white and blue was to be painted on operational aircraft in the Netherlands Indies by 1 March. No pictures of CW-21Bs with these new markings are known to have survived, but there is no reason to assume that the flags were not applied.

So far, the Interceptor pilots had been given strict orders to concentrate on bomber aircraft and avoiding combat with the nimble Japanese fighters. This restriction was lifted on the 24th, after analysis of a downed Ki-43 fighter, which revealed the relatively light armament of one 7.7 mm and one 12.7 mm machinegun. Combat

experience had learned that the CW-21B was a match to the Japanese fighters in one-on-one combat.

Little is known about CW-21B operations during the period between 25 and 27 February, but the situation on Java was growing ever more grim. There were no Japanese air activities on 26 February, but this proved to be calm before the storm; a combined naval force of Allied ships made contact with a powerful Japanese fleet on 27 February. The Japanese ships were escort for a Java-bound invasion fleet. The ensuing Battle of the Java Sea lasted a full day. A significant amount of Allied shipping and lives were lost.

On the day after the confrontation at sea, three CW-21Bs stood standby at Andir, but the Japanese did not show themselves and no flying was done.
With opposition at sea broken, Japanese forces landed on the northern coast of

Java during the night of 28 February to 1 March. Six machines were at the T.D. workshops of Maospati. Of these five were undergoing repairs. A sixth had been stripped for parts to keep other aircraft operational. They were destroyed by own personnel when Maospati was evacuated. As many aircraft as possible were flown out. A single CW-21B was among them, it was flown to Andir.

Four aircraft remained at Andir but availability was limited. Two of the machines were in need of maintenance or engine replacement.
During the morning of 2 March a mission was flown with two available aircraft. Boxman and Haye provided air cover for an advancing army column heading towards Kalidjati airfield, which had been captured by the Japanese. When their task was completed, the two pilots checked the roads for enemy activity. Pilots Boxman and Roumimper flew a second sortie,

this time a strafing attack on the Japanese bridgehead at Eretan Wetan, together with three Brewsters.

A pair of CW-21Bs, led by Boxman, joined four Brewsters during a flight in the late afternoon. They were tasked to intercept a Japanese aircraft formation heading towards Andir. After circling over the designated target area the all clear signal was received.

Upon returning to base, the pilots were warned by ground control for incoming fighters. Indeed, the Dutch pilots were quickly confronted by about 16 enemy fighters. The Interceptor pilots avoided a dogfight with this superior number of fighters and landed without trouble. The fighters were quickly prepared for a new mission.

Immediately after the air raid alarm at Andir was over a new mission had to be prepared. Kalidjati airfield had to be attacked at once. The two airworthy CW-21Bs were part of a mixed formation of four Brewsters and six RAF Hurricanes. The group took off at 18h15, but when they arrived over their target dusk had set in. The fighters made one quick pass over the airfield at low altitude but they were too late to actually notice any activity.

Kalidjati was again attacked on 3 March. One of the CW-21Bs had broken down and had to be repaired, leaving a single operational Interceptor. Due to poor weather several involved pilots aborted their flight. Two Buffalos and the remaining CW-21B reached Kalidjati. Boxman strafed Japanese personnel on the aprons of the airfield.

A single CW-21B was still available when a Japanese bomber force of 52 aircraft plus escorting fighters was heading for Andir. Boxman took off, as did four Buffalo fighters. Boxman was first attacked by a group of A6M fighters, but was also fired at by friendly anti-aircraft artillery. His plane was hit in one of the fuel tanks and caught fire, forcing Boxman to bail out. Three aircraft were at the T.D. workshops during the attack on Andir. One was damaged in the nose area and engine. Another, without engine, was completely destroyed.

P-40ES RUSHED INTO SERVICE

The American aircraft tender *Sea Witch* arrived in the port of Tjilatjap on 28 February, carrying a shipment of 27 Curtiss P-40Es. Twelve of these were selected for service with 2-Vl.G.IV. T.D. personnel worked round the clock to assemble the aircraft, which were spread around the airfield. The first three machines were ready on 7 March. Sergeant Major A.L. Clignett test flew the first P-40E and after receiving instructions 1st Lieutenant Anemaet and sergeant Dejalle flew the other two machines. The engine of one of these aircraft had to be

A P-40E in Japanese markings. There are no photos of intact P-40Es with Dutch markings, but it is assumed they were painted in standard US Air Corps finish, with Dutch flags painted over the American markings. (collection E. Hoogschagen)

A poor, but very interesting photo from a Japanese wartime magazine, showing a captured CW-21B flanked by a CW-22 and B-17 (collection J. Terlouw)

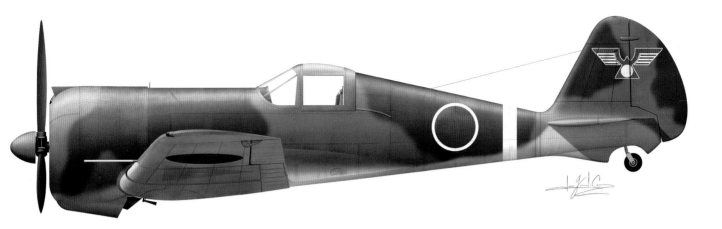

A CW-21B as it appeared at the Tachikawa test centre. Profile art by Luca Canossa.

CAPTURED AIRCRAFT IN JAPANESE SERVICE

examined after Anemaet reported back after a short flight. Two more P-40Es were sent to Andir airfield to be prepared for testing. New orders came in during the evening however. The aircraft had to be destroyed due to the imminent surrender to the Japanese. Seven P-40Es were destroyed using axes and hammers, but five fell into Japanese hands relatively intact. Several other aircraft, stranded on various locations around Java, also fell into Japanese hands.

When the Netherlands Indies capitulated on 9 March two Interceptors were more or less intact and fell into Japanese hands. One had been sabotaged before surrender, but could be repaired. Personnel from 2-Vl.G.IV and the T.D., now prisoners of war, repaired the machine, which was

first test flown by the end of March. It carried serial 1 on the fuselage. It was later transported to Japan for testing at the Tachikawa test centre.

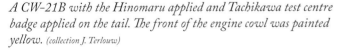

A CW-21B with the Hinomaru applied and Tachikawa test centre badge applied on the tail. The front of the engine cowl was painted yellow. (collection J. Terlouw)

The Japanese repaired a second Interceptor at a later stage and survived the war. It was discovered by British forces at Tengah airfield near Singapore. It was made serviceable and readied for taxi trials, but it was forbidden to take the plane to the air.
(collection J. Terlouw)

ACKNOWLEDGEMENT AND THANKS

I would like to thank the following persons who provided photographs; Thijs Postma, Gerben Tornij, Jan Grisnich, Nico Braas, and Jacob Terlouw. I would like to thank Peter Boer and Max Schep for critically reviewing the manuscript and providing further sources and pictures. Thanks to Luca Canossa for creating the beautiful colour profiles.

References

- Anderson, L., *A history of Chinese Aviation: Encyclopedia of aircraft and aviation in China until 1949*, AHS of ROC, 2008
- Bell, D., *U.S. Air Force Colors 1926–1942, Volume 1*, squadron/signal publications, 1995
- Boer, P.C., *Aircraft of the Netherlands East Indies Army Air Corps in crisis and war times, February 1937 – June 1942*, Batavian Lion International, 2016
- Boer, P.C., *Het verlies van Java: een kwestie van Air Power*, De Bataafsche Leeuw, 2006
- Casius, G., Postma, T. *40 jaar luchtvaart in Indië*, Uitgeverij de Alk bv
- Fausel, R.W., Morley, R.A., *China Odyssey: 1939: A pilot's adventure in the Far East*, American Aviation Historical Society Journal, Fall 1998
- Vries, G., de, Martens, B.J., *Nederlandse vuurwapens – KNIL en Militaire Luchtvaart 1897-1942*, De Bataafsche Leeuw, 1995
- Schep, M.T.A., *De geschiedenis van camouflage en kenmerken op vliegtuigen van de militaire luchtvaart van het Koninklijk Nederlands-Indisch leger*, Uitgeverij Geromy, 2018
- Shores, C., Cull, B., Izawa, Y., *Bloody shambles: The first comprehensive account of air operations over South-East Asia, December 1941–May 1942, volume two*, Grub Street, 1993
- Casius, G., *"The St Louis lightweight"*, Air Enthousiast 16, 1981
- National Dutch Archives, archive 2.10.50.03 - Inventaris van het archief van de Stichting Administratie Indische Pensioenen (SAIP), Stamboekgegevens KNIL-militairen, met Japanse Interneringskaarten, 1942-1996
- *https://www.key.aero/article/popular-misconception*
- *https://www.warbirdforum.com/cw21.htm*
- *www.aviationofjapan.com/2013/04/curtiss-hawk-monoplanes-for-china-pt-2.html*
- *http://digitallibrary.usc.edu/cdm/ref/collection/p15799coll12/id/5796*

Series editor	**Corrections**
Edwin Hoogschagen	Crius Group
Author	**Publisher**
Edwin Hoogschagen	Walburg Pers / Lanasta
Graphic design	
Jantinus Mulder	

First print, January 2023
ISBN 978-94-6456-154-8
e-ISBN 978-94-6456-155-5

NUR 465

Contact Warplane:
jantinusmulder@walburgpers.nl

Lanasta